Worm Farming Business Handbook

A Worm Farmers Guide to Setting Up and
Sustaining a Worm Composting and Vermiculture
System

By

Zera Brooks

Disclaimer

This publication is designed to provide competent and reliable information regarding the subject matter covered. However, the views expressed in this publication are those of the author alone, and should not be taken as expert instruction or professional advice. The reader is responsible for his or her own actions.

The author hereby disclaims any responsibility or liability whatsoever that is incurred from the use or application of the contents of this publication by the purchaser or reader. The purchaser or reader is hereby responsible for his or her own actions.

Table of Contents

Introduction

Worms are often neglected but have been proven to be highly productive organisms over the years. The evolution of agriculture requires lots and lots of manure as the need for good crop harvests is on the increase. Turning kitchen organic waste and food scraps into manure sources to raise healthy crops and thereby creating a major income source sounds too good to be true, but worm farming is a reliable way to do just that.

Worms are creatures that have the propensity to be veracious and capable of consuming food substance that is up to their body weight daily. During food consumption, worms are seen to catalyze food substances, producing natural nutrients for the soil. This helps largely as an organic fertilizer to the soil. They produce a by-product at the end of the recycling process called worm castings (vermicast) which can contain much more nutrients and beneficial bacteria than the original soil and help plants thrive better. Worms have a lot of other ecological benefits like distributing nutrients and decomposing matter, providing protein-rich food for other species like birds and frogs, improving soil structure, and breaking down recycling organic matter within the soil.

Worm farming, on the other hand, is the process of managing and converting organic waste into high nutrient level fertilizers with the aid of these worms. Its profitability is such that it deals with waste products that are zero in cost but yet produce a high demand by-product. This whole idea can be an excellent home-based business while yielding lots of income.

The *Worm Farming Business Handbook* details everything you need to know about starting out as a worm farmer, such as the planning, execution, and maintenance process vis-à-vis the kind of worms to be used, how to get these worms, setting up your bins, how to feed these worms, how to harvest your worm compost and so on.

So, without further ado, let's get right into the business of worm farming.

Chapter 1

The Basics of Worm Farming

What is Worm Farming?

Worm farming (also called vermiculture) is the process of rearing or cultivating worms artificially by converting organic food wastes into nutrient-rich fertilizer, best known as worm manure. Worms automatically begin to feed on the food scraps and other organic matter when you keep them in a bin filled with kitchen waste. As they eat, the worms produce waste known as worm castings (worm poop).

Worm castings, also known as vermicompost or worm manure, are some of the top-tier sources of plant nutrients and minerals with a myriad of valuable micro-organisms that are crucial for disease suppression in plants, root development, and overall plant growth. They also improve the water retention capacity of crops leading to better plant health.

Gardeners and subsistence farmers refer to worm manure as "Black Gold" due to its high nutrient content and the consequent positive effects it has on plants and crops. A regular tablespoon of the "Black Gold" provides sufficient nutrients to feed a small plant for a quarter of the planting year.

Importance of Worm Farming

Your homestead or garden stands to benefit immensely from the cultivation of a worm farm by providing you with a super efficient and self-sufficient way to process and produce organic matter. Having a worm farm gives you access to the most nutrient-filled organic fertilizer while providing a more efficient way to eliminate food waste, as food scraps can be upcycled to run your worm farm better.

Beyond being a form of entertainment for your children, worm farming is developing into a highly lucrative source of secondary or tertiary income as its increasing popularity heralds the spread of a movement across the urban and suburban areas of most western cities. Its popularity can also be attributed to being environment-friendly and low setup cost.

Here are some detailed benefits of worm farming:

Reduction of Greenhouse Gases

If there was ever a reason to consider venturing into vermiculture, prioritizing the good of our environment is one. As food breaks down, methane is produced, which is a deadly greenhouse gas commonly generated in landfills. While we wait for the government and big organizations to do something about our environment, worm farming (or vermiculture) is an individual decision that can be taken to protect our environment; in the spirit of keeping the atmosphere free of greenhouse gases.

Worm farming drastically lowers the volume of food waste disposed of in landfills, which leads to a reduction in generating greenhouse gases because the worms successfully convert organic waste into worm castings.

This also eliminates the need for vehicles running on fossil fuels that are required to collect organic waste.

Protecting The Earth

To the surprise of many, chemical-dependent agriculture has dominated the farming industry, stretching beyond a century, and to the dismay of the life and health of the soil and the people cultivating the land. As a result, the health of the earth and humanity has been adversely affected beyond imagination as humans now have to face up to the reality of a readily diminishing source of food. While advanced innovation in agrotech plays a key role in mitigating the situation, investing directly in the earth through worm composting provides a better solution.

Through the production of nutrient-rich fertilizer and compost from organic waste, the earth we live in benefits through the rejuvenation of its topsoil, leading to the production of quality food crops. With less than 60 years of topsoil left globally, we must take consistent urgent action to prevent catastrophic global famine and prosper in the face of adversity.

Production of Worm Castings

As explained earlier, the waste produced when artificially cultivated earthworms feed on compost is called worm castings. The step-by-step process involved in producing worm castings is known as vermiculture or vermicomposting.

Worm castings are held in high regard and, in some circles, are considered to be the best form of fertilizer in the world. This comes as no surprise as worm castings are rich in nutrients, help in water retention in the soil, improve general drainage and contribute to proper aeration within the soil. The natural biological source of the nutrient-filled soil prevents the toxic substance suffered by other soil types and removes the danger of damaged plant roots.

Minimizing Food Waste

Based on research, it was discovered that approximately seventeen percent of trash in an average landfill was made up of food waste. This is a considerable amount of food wasted, as well as the space used for landfill. The use of the garbage can for food waste disposal has negative impacts on the environment. This is also accompanied by the astronomical costs of transporting the food waste to the landfill. Not only does worm farming reduce the garbage in your bin, but you also help the environment by recycling food and nutrients back into the soil.

Worm Farm Leachate and Worm Tea

Worm tea is produced through soaking worm manure (or worm castings) in water and is a natural liquid fertilizer applied directly to plant roots. There are various methods of making worm tea for local use, which provide the added benefit of preventing plant diseases.

On the other hand, worm seepage or worm leachate is the liquid that drains after collecting at the bottom of your compost bin. Due to the presence of beneficial bacteria present in the worm's digestive system, both worm leachate and worm tea serve as an excellent fertilizer for the soil and plants.

Low Cost of Maintenance

You don't need a degree or an angel investor to start your very own worm farm. And even more amazing is that once you get your operations up and running, you have little to no maintenance to worry about.

As long as you remember to feed the worms (don't overburden them) and keep a close eye on the conditions in the bin, you are good to go. If any problems arise, it merely requires a few minor tweaks to set everything brand new again.

With as low as $300, you can build and successfully run your worm farm from scratch. The low cost of maintenance also ensures that you keep most of the $15,000 to $150,000 yearly revenue that comes from the worm farming business.

Your Kids Will Love It

With the presence of a worm farm, your children can be exposed to an interactive learning experience while giving them an opportunity to cater to the welfare of the environment they live in. They are also taught about the different levels and methods of biological sustainability in a proactive way. They learn from seeing the worms in their natural environment and develop a sense of responsibility as they carry out activities to nurture the worms, such as feeding them and keeping the soil moist. They also get firsthand experience on how gardening clippings and scraps can be recycled to produce something beneficial. Worm farms can also teach your kids the rudiments of self-sufficiency, independence, and autonomy as they carry out the tasks of caring for a worm farm.

The chemical and biological properties of the fertilizer produced from artificially cultivating worms also offer a myriad of learning growth.

Suitable To Be Kept Anywhere

Worm farms are compact, clean, and space-efficient. For instance, you may decide to keep one under your desk at the office, or you may have one in your apartment at home. Even the tiniest spaces like courtyards and balconies can be utilized effectively to grow a profitable worm farm and can be done with minimum effort.

If you are like other people, you may prefer to keep your bin outdoors in the yard as you want to create more living space. This gives the added advantage of putting any unwanted odor or mess out of reach.

The ease with which worm bins can be set up and moved means that you can change their location as the seasons and climates change. It also allows you to check on them as you feel the need to.

They Are Practically Free To Get Started

Depending on the start-up model you are adopting for your worm farm, the cost of starting can be pretty low—whether you're planning to start from scratch with the materials available to you or you're planning to buy a franchise (more expensive) or buying a ready-made mini worm farm.

You can also create a DIY worm bin on your own budget. The equipment you need is not expensive. Some of the stuff you need to get started are food scraps, which are free; you can also make use of ripped-up paper bags, shredded newspaper, or shredded paper as worm bedding.

As a business model, cultivating worms is advantageous because it allows you to start small and grow your worm population per their breeding biology. Worms are hermaphrodites and can perform both male and female reproductive functions.

Worms Can Be Used As Fishing Bait

Composting worms is your best bet if you are going to fish. As a fishing enthusiast, you may already know that composting worms are great as fishing baits. Due to their larger size, the African nightcrawlers are the more popular option for fish bait.

The European nightcrawler is also one of the best because it would continue to squirm and wriggle on the hook for up to thirty minutes after being submerged under water. This constant motion, as well as its reasonable size, makes it attractive to fish and therefore makes fishing more successful.

No Bad Smell

Despite the worm bacterial decay, rotting food, and worm poop in your worm farm (if properly maintained) should produce an earthy smell. Ideally, your worm bin should always smell like a freshly tilled garden. Or even better, it should be completely odor-free.

You may want to consider the setup of your worm farm if you notice any bad smell coming from within the worm bin. If it starts to smell vinegary or rotten, then your worm bin has an acidity problem and needs to be checked.

Make Money

Did you know that you can use an earthworm to make a dollar sign? Yes, you can! But more than fancy wordplay, worms are a good source of income, whether on a small scale with a small worm bin in your home or breeding worms commercially on a large scale.

With a worm farm, you can sell worms as fishing bait to the fishing enthusiasts among your friends, neighbors, or even at the market. Worms also make good pet food for snakes, fish, birds, and a variety of small animals.

The fertilizer your worms produce from composting also makes a great source of income. You can put up the worm castings from your worm farm for sale. It gives you an excellent marketing angle as your worm castings are organic without any pesticides or soil conditioners used to make them.

A lot of people make money from selling worm castings, worm juice (tea), and compost worms. You too can!

Save Money

You reduce or completely eliminate the cost of buying fertilizers, pesticides, and soil conditioners from your budget when you have worm castings or worm tea.

Commercial fertilizers might be good for improving plant growth, but worm castings have the added benefit of enriching your planting soil with the needed nutrients for the continuous growth of plants at practically no extra cost.

Fast Processing and a Quick Turnaround

Of the different methods of composting, one of the biggest reasons starting a worm fan is a favorite is the speed of the composting process. We have a myriad of ways of making manure but maintaining a worm farm is amongst the fastest.

Composting worms can break down organic material pretty quickly. It is common knowledge that worms are capable of eating and digesting half their body weight every day. With this output speed, you should have plenty of fertilizer on your hands.

Types of Worm For Vermicomposting

The type of worm you choose to breed on your worm farm will affect the price it takes to start up, as some worm species are costlier than others. You may find some of the more popular worm species around the price ranges as listed below:

- Red wigglers — $40 to $50 per pound
- Dendrobaenas — Over $50 per pound
- European Nightcrawlers — $40 to $50 per pound

Note: These are the average prices of worms as of the time of writing this book and do not take into account inflation costs. Please visit a nearby worm market for more current prices.

All worm species have their pros and cons. But we have listed the best type of worms for your homestead to help you make a good buying decision. They are guaranteed to help compost your kitchen waste fast and easily. They also come at an affordable price.

The Red Wiggler Worm

This worm has been discovered to be the favorite of many homesteaders. They are ubiquitous in nature, in the markets, and are inexpensive.

In terms of processing, they eat a large portion of food scraps to help with the composting, up to half their body weight. As a species, they are also not picky and would eat almost any kind of food scraps you feed them with. They not only eat large amounts of food scraps, but they can also break down a substantial amount of organic matter within a short time.

Providing them with the right weather conditions (cool and dark) accentuates their breeding instincts as these worms love to mate. The red wiggler worm can also reach lengths ranging from one to five inches and a quarter of an inch across.

The Alabama or Georgia Jumper

This species of worm is native to Taiwan. The *Amynthas Gracilis,* as they are botanically called, works magnificently in composting, especially around tropical and subtropical areas. They are also able to survive in hot temperatures up to ninety degrees Fahrenheit.

The Georgia Jumpers grow larger than the other worm species commonly used for composting and can grow up to six inches in length. Like the Red Wigglers, they are also effective in breaking down food scraps at a rapid rate.

They have a voracious appetite and might need constant feeding as often as every week. Their nimbleness needs to be taken note of as they can easily escape from their bin. They crawl to the surface in search of food and may attempt to jump out of the bins. So if you accidentally leave the lid open, you may not find any Alabama Jumpers when you return.

European Nightcrawlers

Scientifically known as *Eisenia Hortensis,* the European Nightcrawlers share similar traits with the Red Wigglers in that they reproduce easily and are easy to cultivate. But in terms of size, the European

Nightcrawlers are more similar to the Alabama Jumper, which is a larger specie of worms.

The European Nightcrawlers are a great option for fishing and, at their maturity, can weigh as much as 1.5 grams and grow up to four inches in length.

African Nightcrawlers

The African Nightcrawlers (scientifically known as *Eudrilus Euginiae*) are tropical worms like the Alabama Jumpers. Without temperatures above fifty degrees Fahrenheit, these worms will struggle to survive. But if you have your worm farm situated in a place like Florida, Arizona, or Texas, then the African Nightcrawler might be the best choice for you.

But as much as they thrive in hot conditions, they also need enough hydration, so remember to give them enough water and moisture. You may need to feed them every ten days or two weeks as they consume kitchen scraps quickly.

What Are Your Worm Farming Objectives?

In order to choose the composting worm that best fits your needs, you must first determine your big why, the primary cause for venturing into vermicomposting.

Among the most popular reasons are:

- To reduce waste and environmental sustainability
- To make money or generate another income source
- For the sake of growing flowers, plants, and general gardening activities
- To grow worms meant for fishing bait

Worm Farming Terminology

Aeration — The process of letting oxygen into your compost by turning or stirring it.

Aerobic — The aerobic process requires oxygen. In a compost bin, aerobic conditions are preferred. The presence of oxygen is required for aerobic organisms to carry out their life functions. Proper aeration within the bin will prevent unpleasant smells.

Anaerobic — A process that does not require oxygen. When the conditions inside the bin are anaerobic, there will be unpleasant smells rather than the preferred earthy smell. Anaerobic organisms do not need oxygen to survive.

Bedding — The leaves and newspapers are sometimes used as a medium to aid worm composting.

Browns — Composite materials that have high carbon content. They are typically dry in nature

Cocoon — Egg cases that can contain between two to twenty worm eggs.

Compost — This is the final product manufactured after the entire composting process. Also referred to as humus, compost is a dark, nutrient-filled soil conditioner produced through the breakdown of organic matter.

Composter — A box, bin, or any other similar container used for worm composting

Compost tea — The water containing the micro-organisms and nutrients of the compost after it has been steeped to create plant fertilizer in liquid form.

Compostable materials — Materials which are organic in nature and are added to a compost bin for decomposition.

Critters (also compost) — The macro and micro-organisms which are part of humus and help in breaking down organic matter.

Composting — The natural biological process of creating fertilizer and humus through the breakdown of organic matter such as food scraps and leaves by decomposing organisms for the enrichment of plants and soils.

Decay — To decompose, rot, or break down.

Decomposition — A natural phenomenon needed for plant growth which could also be made to occur in controlled environments like worm bins. It happens when dead organic matter is broken down into its basic elements of inorganic or organic matter.

Ecosystem — A complex system made up of interdependent elements like plants, animals, weather conditions, and other organisms that come together to form a cycle of life.

Fertilizer — A substance (natural or man-made) made up of different chemical elements which are used to provide nutrients for plants and enrich the soil.

Food scraps — Refers to the vegetable scraps, uncooked fruits, and other unavoidable waste products used as feed in the cultivation of worms.

Greens — Compost materials that have high nitrogen composition (usually wet)

Heap — An unenclosed pile of plant materials left to decay for the sake of composting.

Humus — The final product formed after the completion of compost by the decay of animal and plant matter, such as leaves. Humus acts as a reservoir for plants by providing a constant source of energy for micro-organisms, improving the water holding capacity of the soil and improving the overall soil friability and soil structure.

Leachate — The liquid containing materials that have been suspended, dissolved, or removed through the breakdown of solid waste in worm castings. The leachate produced from a worm bin or compost bin is a rich source of nutrients and serves as a better soil fertilizer.

Leaf mold — Created by allowing leaves to sit and slowly decompose with time.

Macro organisms — Organisms that are visible to the human eye

Micro-organisms — Organisms of micro size that cannot be seen without the use of a microscope.

Mulch — A layer of partially decomposed plant materials, compost, or tree bark placed around plants and shrubs or over garden beds to serve as protection and enrichment for the soil.

Organic matter — Any matter containing carbon compounds that are or were once living or previously formed by living organisms.

Overload — The oversupply of food items like kitchen scraps into a worm bin which impeded the aerobic processes.

Red worm — a species of earthworms suitable for vermicomposting. The red wiggler is a red worm that is a favorite for homesteaders looking to begin worm farming.

Rodent resistant — The design, modification, or reinforcement of a compost bin to protect it from pest invasion.

Screening — The process of sifting humus to free it from clumps of uncomposted matter and provide a better planting experience.

Soil — The upper layer of the earth, which is made up of sand, clay, silt, tiny rocks, as well as decomposers, and organic matter for the growing of plants.

Soil conditioner — Also known as soil amendments, are added to the soil to increase its organic content and enrich the physical condition of the soil.

Vermicompost — Refers to the process of performing worm composting using various worm species or the by-product produced by composting with worms. Worm cocoons, worms, broken down organic matter, and worm castings are some of the components of vermicompost.

Vermicomposter — A person who does worm composting or a bin/box used for composts.

Vermicomposting — Composting with worms

Vermiculture — This is the artificial process of cultivating worms to create nutrient-rich soil using food scraps.

Worm bin — A vermicomposting system or a specially designed container manufactured for worms to eat organic garbage and to live in

Worm castings — Also called poop or worm manure, is the waste produced by worms that serve as great substitutes for commercially sold fertilizers.

Wet garbage — This usually refers to compostable organic materials, garden waste, grass clippings, and food scraps.

Chapter 2

Planning for Worm Farming

Choosing Your Worm Farm

Worm farms are available for purchase, as earlier mentioned, or you can construct your own worm farm (discussed shortly) in various sizes and forms to fit your taste and needs. Worm farms typically take up very little spot and are made up of stacked trays of bins with leg support of some sort. They are the perfect dimension for a small family. Should you make up your mind to set up farm worms with more processing power, one that can handle enormous amounts of food scraps, then you can build one out of a recycled bathtub or one procured from industrial wheelie bin worm farms. Larger families, schools, cafes, restaurants, or workplaces are good candidates for these larger worm farms because they produce a lot of food waste. It's crucial to pick a farming system that is ideal for the space you have and can handle the amount of waste produced.

Also, it's crucial to take into account the worm farm's "footprint," or spaces occupied on the ground. From the above image, all three of my work farm take up some space outwardly and are fortunately located in a secluded portion of my backyard. It wasn't intended for me to have all three farms simultaneously. I started with one, adding one more to consume crumbs from the kitchen. Worm farms made with bathtubs have 200-liters volume and take up the same amount of area as a bathtub, which is quite a bit! Worm farms made from wheelie bins consume very little ground space with 140-liters volume, 240 liters, or 360 liters. You can benefit from it being that it is mobile due to its wheel design.

Organic gardeners can take advantage of worm farming as a good investment, plus it lasts long. So take your needs into consideration before constructing your own bin or making a purchase.

Getting Your Worm

The red wiggler is the kind of worm that is most frequently employed in vermicomposting. Because of this, this vermicomposting technique is also known as red worm composting. The red wiggler (Eisenia fetida) species thrives in decaying organic matter and manure and is incredibly effective at decomposing it. They do not bother to live in a small bin and wouldn't try to dig down further compared to your regular garden earthworm.

Red wigglers were previously only sold at fishing bait shops, but because of their current level of popularity, several suppliers can now deliver them to you by the bag.

When you gather your food scraps and have them weighed, the amount of worms needed can be determined because a pound of worms will consume half their body weight in food scraps each day. One pound of worms should be greater than two big handfuls and contains roughly 1000 worms. Without knowing the amount of worms to be collected, your worm bin may not function properly. My worm bins twice experienced fly infestations that made the bins unsafe. I was baffled as to why. Due to my lack of knowledge of what one pound of worms should look like, it emerged that two separate companies significantly understated the amount of worms I ordered. I fed the worms way too much as a consequence, and the waste decomposed and supported the growth of flies.

Here are some useful worm-finding suggestions, from small-town breeders to large businesses!

Identify regional worm vendors

It's a good idea to help out regional or local worm vendors wherever we can. Although purchasing worms online can be more convenient than going to a small farm, a lot of small businesses now sell worms online, and most of them provide express shipping.

Sellers of worms in your state

You can use this directory (https://bit.ly/3QX5MQQ) to look for online worm sellers in your region! Even though you may be unable to pick up the worms in person, the worms, if they can travel more quickly by mail, will be happier when they arrive.

Well-known small businesses

You couldn't locate worm sellers in the directory, or did you? If you didn't, here is a list of the top online worm sellers we have put together. Since these businesses are frequently owned by a single person, some of their websites may appear a little shoddy, but that simply adds to their charm! They all have positive online reviews from the gardening and composting communities.

Worms from Uncle Jim's (https://bit.ly/3Crfqam)

Texas Worm Farm (https://bit.ly/3dNK57q)

Worms etc (https://bit.ly/3dNDPgb)

Worms from TerraVesco (https://bit.ly/3PAnwAB)

Delta Worms (https://bit.ly/3wnsxpb)

Midwest Worms (https://bit.ly/3dNDPgb)

The Worm Farm (https://bit.ly/3AfC3vD)

Worms from the Meme's (https://bit.ly/3KdZx8Y)

Kijiji, Facebook Marketplace, and Craigslist

Would you rather obtain your worms directly? On Craigslist, anything may be found. See what comes up by performing a search for "compost worms" or "vermicompost" on any online marketplace (such as Craigslist, Facebook Marketplace, or Kijiji).

Large Retailers That Sell Worms

Home Center

Many people turn to Home Depot for their gardening needs. Yes, compost worms are probably available at your neighborhood store, but we advise phoning ahead to avoid having to make the trip if they're out of stock.

Walmart

Sometimes, during the height of the gardening season, your neighborhood Walmart will have compost worms in stock! However, not all stores will have them in stock, so once more, call ahead before visiting.

Landscapers and nurseries

Like Walmart, not all nurseries and landscapers have worms, but if one is close to you, it's worth calling.

Setting Up Your Worm Bin

Any container that resembles a box, most frequently, can be considered a bin (can be used as indoor or outdoor worm bins). Most often, materials like plastic and wood are used to shape them (which can be of lumber or plywood material). You can create a temporary worm bin or composter from old household materials like a repurposed garbage can, a broken refrigerator, a malfunctioning toilet, or an old wooden drawer.

Worm composters have perforations that allow room for ventilation and moisture. It is mostly made of plastic and hoisted above ground level to allow water to flow freely to its bottom.

Worm composters or bins (sometimes called vermicomposters) are typically affordable and simple to maintain. There are various approaches to vermicomposting. Worm composting bins can also be bought. Because you don't want the worms freezing in winter or getting too hot in summer, you should place your bin indoors. Additionally, because the composter will be creating compost and worm "tea," you might wish to place the bin in a basement or another inconspicuous location.

What You Need

The building instructions for an indoor worm composting bin are discussed below. The following items should be purchased, borrowed, or repurposed to begin worm composting:

1. Two plastic containers: The smaller container must fit inside the taller container.

   - There is no need for the shorter (bottom) bin to have a top. It works wonderfully using a rubber or plastic container that is 15 inches deep, 25 inches wide, and 5 inches high. The length allows you to remove the excess liquid, or "worm tea," to be used in places like your backyard or for vegetation.

   - The worms must be prevented from escaping the box; hence, the top of the tub needs to have a lid. For you to be able to drill holes in it, it must also be flexible. It works nicely to have a 15-inch deep, 20-inch broad, and 15-inch tall 18-gallon tub.

Why holes?

Holes allow the inflow of air when drilled close to the tub's top, thereby making it possible for worms to breathe. The worms are prevented from drowning when holes are drilled close to the tub's bottom, allowing excess water to drain out of the box. A bin with holes makes it simpler to maintain the bin and control the liquid; however, you can manually control the liquid, but it might be challenging to get it right. A thin vinyl screen is placed over both sets of openings (holes) to prevent the worms from escaping.

2. A drill: A one-inch diameter drill and a drill bit with a diameter of one-eighth inch are required to bore the aforementioned holes.

3. Screening material: For this, you can use window screen material. However, avoid using metal because it will rust when it comes into contact with the moisture from the bin. There are only four screen strips required, each measuring 4 by 4. As mentioned earlier, the worms can escape if the

openings are left uncovered, hence the need for a screen.

4. Water-resistant glue: Keeps the screen firm, even when they are wet.

5. Bedding material (e.g. shredded paper): Fill your bin three inches deep with shredded paper and add extra whenever you feed the worms (typically weekly). Any type of paper will do but stay away from glossy, hefty colored paper.

   Other bedding options are:

   - Egg cartons
   - Leaves
   - Straws, and;
   - Shredded cardboard

   I advise leaving the straws and leaves outside because they may attract critters and insects.

6. Some dirt: One pound of dirt should be sufficient. Always ensure it is free of any dangerous substances. If it goes according to plan, the worms will soon start making their own compost.

7. A little water: Water moisturizes the dirt and paper to create an environment that is conducive to the worms' growth. Before usage, the paper needs to be soaked and then drained.

8. Worms: Kindly review the types of worms section earlier discussed to decide on which worm best suits your purpose. However, red wriggler worms are recommended since they consume garbage quickly; they should weigh one pound. The Asian Jumping worm is invasive and sometimes sold as Alabama Jumper or Georgia Jumper, so kindly take note of this.

9. A trowel: Required to move the compost when necessary within the bin.

10. Container for food scraps: Collect any leftover fruit or vegetable pieces in a container with a tight-fitting lid. Why not just dump the food in the worm container instead? This is because worms develop well when fed once a week. After all, they thrive when left unattended.

Preparing The Bins

The actions to follow to get the bins ready are given below:

1. On the side of the taller container, drill a hole of about 1-inch (roughly about 2-inches from the top). On the other side, drill a second hole. At the corners of the bin's bottom, drill four 1/8-inch holes.
2. Vinyl screening should be used to cover each hole, and the water-resistant adhesive should be used to secure the screening. Before moving on to the next stage, ensure that the adhesive is totally dry.
3. Put the tall container inside the smaller one. Ensure you do not make any holes in the short bin.

Finishing Your Setup

To get the worms off to a healthy start, you must moisten the bedding materials because they cannot thrive in a dry environment. Start pouring water into the clean bucket or tub where you've placed your bedding materials. The ideal water for your worms is dechlorinated water.

Tip: Load a few cups with water from your tap and set them outdoors for one day or two to dechlorinate the water. The water's chlorine will dissolve, leaving your water chlorine-free.

Then simply place the bedding in a clean bucket or tub and begin pouring in the dechlorinated water to saturate it. A little at a time, add water to the bedding and thoroughly combine. Your bedding material should have the texture of a wrung-out sponge, i.e., quite moist but not soggy. Squeezing a handful of the bedding should cause a few drops to fall out; if more water comes out, add a little extra dry bedding to balance the moisture level.

Then, mix in a shovelful of completed compost or garden soil (dirt) with your bedding material. It gives the worms something to eat and adds micro-organisms to help the materials in your worm bin break down faster. Fluff up the bedding a little before throwing it into the tall bin (the mixture should be up to three inches in depth). You want your worms to be able to move around the bedding seamlessly. Large bedding clumps should be broken apart.

Bins should be solid colors not clear

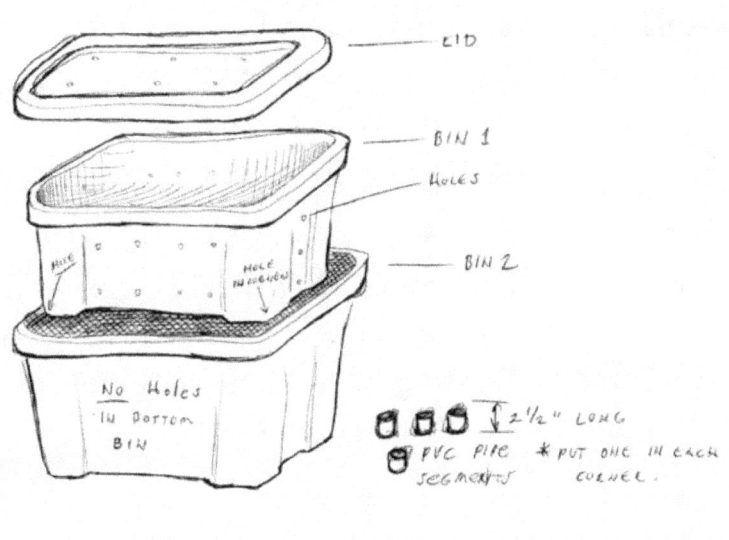

Where To Put The Worm Bin?

Worm bins can be placed inside or outside, as mentioned earlier, but both have advantages and disadvantages that apply to any climate. There are two major advantages to having your bin outside: capacity and far less focus on details.

It won't really matter if your worm bin stinks when kept outside because you won't be bothered about it that much. However, if you don't live in an area with much warmth, the worms will freeze to death, or you'll need to take them inside.

Even within your home, worms can be kept in an indoor bin, but with limited capacity because you will always be constrained as you can only have a certain amount of worms eating just some amount of food before things start to smell and go bad.

Look at this design (https://bit.ly/3CkqWnR) from Nature's Always Right, an urban farm in Lemon Grove, California, if you are considering constructing a higher capacity outdoor worm bin system that is suited for a bigger home garden or urban farm.

Feeding Your Worms

Worms will consume practically any leftover fruit or vegetable food. Nothing from an animal, including dairy, fats, bones, or meat, should be given to them. They struggle with certain stems, onion's outer layers, and an excessive amount of citrus. Ensure sure that the amount of citrus waste added does not exceed 1/5 of the overall waste if you have lots of orange peels or any other citrus. Never offer them any carnivore feces, including dog, cat, or other carnivore feces. Worms won't consume plastic coffee filters, tea bags, or the labels that supermarket retailers put on food. However, you can also include non-plastic teabags, napkins, shredded paper towels, and coffee grounds.

Dig a hole and lay the food in your bin under the bedding's top layer. Breaking up bigger chunks will greatly help. Also, check to see if there is anything from past feeds that the worms are shunning before adding a few handfuls of shred paper. Take away anything that isn't edible.

As previously stated, worms will consume half of their body weight each day. Nevertheless, they will start out a bit slow when you initially acquire them. Start off with little portions of food and monitor how long it takes them to finish it. Once the equivalence of half of their weight is added daily, then increase the quantity.

Reduce the size of the food you are providing if they appear to discard a lot of it. Practically, one pound of worms should consume three pounds minimum each week and twelve pounds of waste each month.

After the food previously introduced to the worm bin has been consumed, then you can introduce more food.

A few things to note;

- Food scraps should be taken to the worm bin.
- Using a trowel, gently make a hole for the scraps to be introduced into.
- Food scraps should be completely covered with dirt and wet paper. Fruit flies are drawn to exposed food but not to food scraps that are covered. Dirt and wet paper should be added to the bin until there's sufficient compost created by the worm to cover the food scraps.
- Observe the things the worms eat and what they aren't eating. If your worms haven't eaten any scraps in a while, remove them because they

might not enjoy them. Reduce the volume of the scraps if they are too big.

- Put the worm bin's lid back on.
- The food scrap bin has to be cleaned for the upcoming week.

Troubleshooting Worm Bin Issues

Sticking to the recommended amounts when adding food to the bin will help you avoid multiple issues. The following are solutions to typical worm bin problems:

Fruit flies in your worm bin: You are well aware of how simple it is to have a fruit fly infestation if you've got food on your kitchen counter. The key is to prevent it from happening. Be careful to thoroughly cover the food scraps with the bedding, chop it up into small bits, and avoid allowing anything to sit and fester in the bin.

Your composter smells bad: If you notice that the compost is starting to stink, burrow the bedding to take out food crumbs that are decomposing. The worms may leave something to soften up, even though they are meant to be consuming everything before it becomes moldy. For instance, they generally won't consume the entire fresh skin of banana before it begins to degrade and melt. They will process food more quickly if you cut up larger bits. An unpleasant odor can also be an indication of a moisture issue. The bin will become saturated and quite moisturized for the worms if there is excessive waste rot present. To help restore equilibrium and bring the bedding's moisture level to a porous state, add a little paper bedding and dry cardboard to the bin.

Worms are emerging from the bin: A few worms escaping the bin is typical, but if there are too many, then there may be an environmental issue, such as humidity, temperature, and perhaps even pest problems. Increase the moist level if the bin is arid. Add some more bedding if it's too moist. Examine under the bedding's top layer for likely the issue or inspect for insect larvae.

There is a pest problem: Other creatures occasionally adore the dark, damp atmosphere of the bin—even only for egg-laying. Place the bin outdoors and collect the castings if you see any bugs, eggs, or even fuzzy critters. Now is an excellent time to clear out the bin and dispose of expired food or worn-out bedding. Restart afresh, putting the worms back in their dwelling place (home). Use cords or boulders to secure the lid if you notice any rodent intruder.

If troubleshooting doesn't solve your issue, then try to change the bedding with any of the available amendment options below:

Material	Purpose
Coconut coir	Balances out nitrogen-rich materials like kitchen scraps and grass clippings, controls moisture, and improves quality of worm castings.
Glacial rock dust	Enhances the microbial population, increases nutrients, and improves water retention.
Pumice	Increases air circulation, prevents matting, and controls moisture.

Maintaining Your Worm Bin

Worms are fortunately rather low-maintenance animals. Your worms will prosper as far as the surroundings are ideal. Now that you've made the decision to start your own worm farm, you even bought (or constructed) a new habitat for them, and everything is set up and prepared for your worms to start producing the much-needed compost, the next ideal thing to do is to ensure that you continuously sustain your worm farm through consistent maintenance practices to ensure your desired outcome is achieved. So long the conditions are right, your worms would thrive.

Here are eight points to remember while you learn the rudiments of maintaining a worm farm, some of which we have covered previously.

1. Make sure there is proper drainage in your worm bin

The openings at the bottom can become clogged very easily. It's also crucial for the drainage holes of your worm bin's bottom to be kept free if it's outside in the rain to prevent flooding. Additional drain holes should be added if you are unsure. If the bin was created from a storage container, more holes should be added to the middle of the bin because it could sag after extended use.

If you don't notice any liquid dropping from the container's bottom, you should absolutely ensure the water isn't sitting in the container.

2. Your worm bin shouldn't be wet but moist instead

Your worm bin's moisture level must be precisely correct. Worms have skin-based respiration. The worms will grow dehydrated and perish if the environment is too dry. They risk drowning if the water is too wet since they won't be able to breathe. In addition, when the worm bin is very wet, the worms become anaerobic; this is unhealthy for the worms.

3. Do not feed your worms too much

The most common error people with their first bin commit is this. Although worms do consume their body weight each day, a sizable percentage of that is likely to be bedding. You risk killing your worms if excess food is added, particularly in a small bin. The food has the potential to heat up, increasing the bin's temperature and making your worms uncomfortable. The bin might also get acidic once more, which would be unpleasant for your worms. As a general guideline, avoid adding extra food than your worms can eat in a month.

4. Add bedding each time you feed your worms

There is never enough bedding. It's an excellent practice to get into the routine of adding extra bedding, laying the food on top, and then putting some more bedding to cover it all. The water that is released as the food decomposes will be absorbed by the bedding at the bottom. Fruit flies won't be drawn to the food in your bin because of the bedding on top.

5. Maintain adequate airflow in your worm bin

This is crucial when relocating your worms for the first time. Your worms may need some time to adjust to their new surroundings after being moved. It might be beneficial in this situation to leave a fan running for the first few nights.

If you bought a worm farm, it's a good idea to leave a few empty trays on to allow the system to circulate more air. Try boring some holes in the top of the empty trays if the worms appear to be climbing up into them. This will help the air circulate.

6. Your worm bin's airholes should be sealed

The air slots should allow air through while keeping insects out. Adding a mosquito net to your bin's inside is an excellent technique to let air in while keeping insects out if you are utilizing big air holes. As discussed much earlier, you can also use a thin vinyl screen to seal the holes.

7. Your worm bin should be kept away from light rays

Earthworms don't enjoy lots of light. This is why transparent worm bins are never sold, and if you come across one, don't buy one. Your worms will be happier if your bin is kept away from direct sunlight, such as in a cupboard, under the sink, or garage.

8. The worm castings should be harvested when ready

Worms are no different from other animals in that they do not wish to spend the entire day crawling around in their own waste. Worms in tiered bins will, for this reason, go from the bottom tray, which is filled with worm castings, to a tray that is mostly filled with food and fluids. It's a little cruel to let your worms swirl about in their own feces. Don't hesitate to harvest your castings once they are ready. Just be sure the compost is harvested (typically at the end of the week) before feeding the worms once more. More on this to be discussed deeply in the next chapter.

A Short message from the Author:

Hey, I hope you are enjoying the book? I would love to hear your thoughts!

Many readers do not know how hard reviews are to come by and how much they help an author.

I would be incredibly grateful if you could take just 60 seconds to write a short review on Amazon, even if it is a few sentences!

>> Click here to leave a quick review

Thanks for the time taken to share your thoughts!

Chapter 3

Harvesting Your Worm Compost

You must frequently harvest your worm castings if you have a vermicomposting system. This "black gold" is the ideal fertilizer for your plants because it is nutrient-rich. The worms' health is also maintained through harvesting. When most of your worm bedding has turned into a rich, dark mixture of vermicompost and worm droppings that looks like soil, it is time to harvest.

Worm castings can be collected in a number of effective ways for small home-based worm farms. Your preference and the type of system you have set up will determine the approach you would take.

There are stackable multi-tray systems, single-tray vermicomposting systems, and numerous DIY systems. Pick a technique that removes the casting effectively while leaving enough worms for the compost to continue to break down.

Method 1: Promote Worm Relocation

Worm relocation is a relatively common harvesting technique that relies on the idea that worms will move toward food. If using a single tray system, collect any uncomposted scraps still in the castings and move them to the middle or to one side of the tray. Only add fresh food where you want the worms to be.

Food should be placed in the tray above the one you want to harvest in a multi-tray vermicomposting system. The majority of the worms will move spontaneously to the food source in one to four weeks, depending on the size of the tray. You will then have almost worm-free castings to pick as a result. You can also fish out any remaining eggs at this time. These little yellow eggs will hatch into worms later on. Anyone who is uncomfortable using their hands to handle worms will find this method easy to implement.

Method 2: Relocation Using Light

Applying artificial light or sunlight to your castings can also result in worm-free castings because worms will naturally flee from light. Avoid exposing your worms to too much heat or sunlight because doing so could cause them to dry up and perish.

One of two options is worth trying:

- Build casting mounds on a clean, solid surface after gently emptying your worm tray. The worms will tunnel to the bottom and away from the sides in about twenty minutes to avoid the light. The tops and sides of the pile can then be scraped off, working your way down until only small piles are left.

Or

- Gently transfer the worm castings to a container for short-term storage. The now-empty worm tray should be refilled with food and bedding. The worm tray should be covered with a piece of burlap or another material that has holes the size of the worms. Wait 20 minutes before spreading a 1-2 inch thick layer of casts and worms over the cloth. The worms will have descended into the worm tray by that time. Apply the worm-free castings to the soil of your plants (if you so choose) after placing them in a storage container.

Note: You can make your own filter trays out of wire mesh rather than burlap. Implement this procedure as soon as possible for the worms to stay damp. Worms can get dried out by both sunlight and human hands, which is harmful.

Method 3: Harvesting By Hand

This approach may be preferable for those who are willing to move at a slower, more intimate pace or who only require a small number of castings at the moment. The process is the simplest of all; all that is needed is to gather a few handfuls of castings and sift through them for worms, which can then be returned to the tray. You can empty the entire tray at once, fill it up with food bedding, and then sort through everything. If the time is not right for a full harvest, you can also choose only a few handfuls. This kind of sorting is frequently entertaining for kids.

Method 4: Screen Composting

You can simply remove the worms, sticks, and undigested debris from the completed compost by sifting the worm bedding through a screen. Compost that has been screened is particularly prized for being fluffy, light, and debris-free. A composting screen can be made or purchased.

See the compost screening instructions below.

Things you need for screen composting

- wood to construct a frame for the screen and;
- a ¼" or 1/8" hardware cloth for the screen

- a compost bin or pile with lots of time to enable it convert to fertilizer
- a place where you can shake your compost (a tote, a cardboard piece, or a clear area on the pavement or ground)
- a trowel or tiny shovel

Steps

1. Verify the bin's moisture content. It should resemble a wrung-out sponge in texture. (If you have a

moisture-measuring instrument, it should register a moisture content of no more than 80%.) If it is too wet, thoroughly stir in some coconut coir. Sit tight for a day or two for it to start to dry out. The compost will obstinately cluster into oval chunks that resemble deer droppings above the screen if it is overly damp.

2. Put the screen above the intended bin.

3. Add some compost that has been produced to the screen.

4. Side to side, shake the screen. Large items and worms will sit above the screen while the compost will fall through.

5. Set the worms and bigger items away. Hold worms inside and away from the sun since they can dry out rapidly or attempt to flee.

6. Steps 3, 4, and 5 should be repeated until you've sufficient compost

7. Some bedding (at least an inch in depth) should be left for the composting worms. Create more worm bedding when needed.
8. Cleanup the rubbish and worm pile. Take away inedible objects like twigs, pips, stones, and stickers. Reintroduce the worms and uneaten food into the worm bin. Allow the worms to dig down on their own to avoid suffocation. They'll dig deeper if you shine a light on the bin for a while. Feed them.

Once you've collected your worm castings, you can keep them in a clean bucket, bag, or other container until you're ready to use them. They are teeming with nutrients and beneficial soil micro-organisms that are ideal for adding to the soil. Worm castings should be collected on a regular basis to keep the worm population healthy.

Where can my worm compost be used?

Worm compost can be applied to the soil's surface or mixed in with soil around plants, just like any other high-nutrient fertilizer, although be cautious not to add too much as it might burn fragile stems. Worm compost tea (or worm leachate) that drips from the bin's bottom is very concentrated. To avoid any form of burning when using it to enhance the soil around your plants, dilute this tea using a ratio of 1 part leachate to 10 parts water.

Chapter 4

Worm Farming FAQs

What is the worm's life cycle?

A worm's life cycle starts as an egg inside a cocoon. The offspring emerge from this cocoon at around 4 weeks after it forms, and it is less than the size of a rice grain, lemon-shaped, and yellow in color. They are known as "juveniles" as they hatch, are little longer than 12" long, and have already begun to consume organic material. The juvenile worm matures into a "mature" or "adult" worm after roughly 40 to 60 days. Although worms are hermaphrodites (containing both male and female reproductive organs), they still require another worm in order to mate and reproduce. This is because their clitellum has developed and contains their reproductive organ. Once the process is over, both will expel an egg, starting a new life cycle.

How are worms able to grind their food?

Because worms can only fit microscopic particles in their mouths, microbes will soften food materials before worms try to ingest them. Composting worms like Red

Wiggler Worms and European Night Crawlers have powerful gizzards. To help with the grinding, their food is combined with substances like sand, limestone, or soil. The food item is broken up into tiny bits as the gizzard muscles contract and mix with the worm's bodily fluids.

How long do worms live?

Due to their habitat, all worms have varying lifespans. Most worms in the wild live and die in the same year because of things they can't control, like food, dryness, and weather that is too hot or too cold.

In a controlled setting, Red Wiggler worms (Eisenia fetida) can survive for up to four years. In a controlled setting, the European Night Crawler (Eisena hortensis), can live for up to two years.

In the worm bin, do worms perish?

They will, unfortunately, perish in the worm bin because it is a natural part of the cycle. They are difficult to locate because, with up to 90% water in their bodies, they decay so swiftly. If you do discover dead worms, you need to identify the issue and fix it (too hot, too wet, too acidic, wrong peat moss). The issue can occasionally be resolved by switching out the bedding.

How can I determine whether my worms are reproducing?

Red worm eggs resemble little lemons with a straw tint. Even though they are much smaller and don't have a bright red color, baby red worms look the same as adult red worms.

Are worms in need of air?

Yes, worms require oxygen to survive; after all, they are living things. From the outside to the inside of the worm, air or oxygen diffuses across the moist skin tissue. The carbon dioxide created by the worm's physiological processes travels from the inside to the outside of the worm in reverse. The bedding must have a consistent flow of fresh air.

How do my worms fare over the winter?

Your worms' movement, feeding, and reproduction rates will increase when the temperature drops, depending on where they are kept (shed, garden, garage, house, etc.). To avoid creating a composting situation due to uneaten food in the worm bin, be sure to watch the amount you feed and the length of time it takes for them to eat. Worms do not digest as much food waste in the cold. Depending on where you live, this period often spans the months of November to February.

Why is there growth in my worm bin?

In your worm bin, molds, fungi, and/or sprouts may occasionally appear. These are a normal part of the composting process because they aid in the breakdown of food waste. The worms consume both of these.

What additional types of species are present in my worm bin?

White worms, springtails, and millipedes are the most frequent composting companions. Your worms won't be harmed by them. There is a possibility that centipedes will get into the bin and endanger your worms. Remove them as you find them; if you find a lot of them, you'll need to replace your bedding. Millipedes have two pairs of legs on each segment, compared to centipedes' one pair.

Will common earthworms be effective for composting?

No, common earthworms won't work for composting; if introduced to an indoor worm bin, they will perish. Red worms, which are also called composting worms or European night crawlers, live on the surface and don't burrow. They like to live in organic waste that is high in nutrients, like manure, compost piles, or leaf litter.

What should I do with the worm tea?

Your worm farm's liquid output is called worm tea (or worm juice). This substance is frequently called "liquid gold." Use it as a liquid fertilizer on your plants after diluting it with water.

How soon will my compost be completed?

I've read that a worm farm level can be completed in 3 to 4 months. How big everything is and what you have put in there will determine the length. With the exception of a few huge egg shells and avocado skins, the bottom level of my worm farm has been set for a few weeks (about 3 ½ months).

Are eggshells beneficial to worms?

Eggshells are beneficial to worms, and you should put them in your worm bin for a number of reasons.

Eggshells supply calcium, which lowers bin acidity and supports the reproduction of worms, thus ensuring a healthy worm population.

Is worm farming profitable?

Yes, raising worms can be both a simple and highly lucrative business from a financial and personal perspective.

What is the potential profit from selling worms?

Worms reportedly retail for between $30 and $32 per pound. Particularly, if sold as retail in large volume, a successful company will sell 350–400 worms each pound, making roughly $0.08 per worm.

How much money are worm farming businesses making?

Businesses that raise worms make between $15,000 and $150,000 annually. When learning initially and starting up, this might appear to be nothing more than a side hustle, but just like with most businesses, it will take time to establish your company's structure and clients.

How much can red wiggler worms be sold for?

As per sources, you can earn up to $65 per pound (or around 7 per worm) if sold on eBay. For your worms to be sold independently and to set your own prices, you can open a Shopify store. For example, buckeyeorganics.net charges roughly $36.95 for each pound.

How long can a worm farm be left unsupervised?

In general, if you provide sufficient supplies and ensure the environment is good for them, a worm farm can be left unsupervised for up to 6 weeks.

The end... almost!

Hey! We've made it to the final chapter of this book, and I hope you've enjoyed it so far.

If you have not done so yet, I would be incredibly thankful if you could take just a minute to leave a quick review on Amazon

Reviews are not easy to come by, and as an independent author with a little marketing budget, I rely on you, my readers, to leave a short review on Amazon.

Even if it is just a sentence or two!

So if you really enjoyed this book, please...

\>> Click here to leave a brief review on Amazon.

I truly appreciate your effort to leave your review, as it truly makes a huge difference.

Chapter 5

Profiting From Worm Farming

You have the opportunity to start a worm farm and become environmentally conscious while also making a good living. Many people involved in worm farming now initially saw it as a hobby or pleasure; they never thought it was worthwhile as a profession.

The business had a relatively small number of entrants at first because they were just raising worms to sell to bait stores for small-scale fishing. However, as environmental sustainability has become a worldwide concern, worm farming has become more and more well-known.

The low initial investment is a big part of what makes worm farming so attractive. The business can be launched from your backyard, an empty chicken coop, or even your garden. You can get started right away and earn money almost immediately if done right.

This chapter focuses on taking your worm farm passion and transforming it into a business-making enterprise, which piggy's back on most of what has been covered in preceding chapters.

So, let's get right into it.

How to Start a Profitable Worm Farming Business in 23 Steps

1. Have an understanding of the market

Although earthworms are sold all year long, worm farmers are busiest in the spring and summer. Worm farming is a simple way to turn leftover fruit and vegetable waste into high-quality potting soil or soil amendment for your garden or indoor plants.

Both homeowners and apartment dwellers are able to do it throughout the year. People who want to compost their food scraps but do not have space for a backyard compost bin will find worm farming to be especially helpful. Since there are no longer any industry standards or market rules, the quality of products varies a lot.

As a result of increased buying costs, soil degradation, poor plant and soil health, and consumer preferences for non-chemically cultivated produce, more and more farmers are today abandoning chemical fertilizers. Due to the increased demand for organically cultivated food, several opportunities have been created for those who produce vermiculture by-products to enter a market that is expanding.

When farmers use vermiculture and its by-products in their farming, they can expect to gain in a number of ways, such as better soil fertility, better plant and animal health, higher yields, and lower costs of production.

The worm farming business is well suited to be run by a team of husband and wife in most circumstances, even with only a spare acre or less. Obviously, this relies on the goals you have in mind as well as the amount of organic wastes you want to transform into compost.

2. Conduct market and feasibility research

- Statistical and psychological data

The demographic and psychographic makeup of people in need of worms includes members of all social classes, the public and private sectors, homes, communities, and individuals from all walks of life. Knowing the

74

kinds of customers who will require a worm farm's products and services is crucial.

The people and organizations in need of worms are listed below.

- Farmers grow crops
- Fishermen's and Farmers' Research Institute
- Aquatic farmers
- Universities
- Manufacturers of organic products
- Laboratories and;
- Producers of animal feed

3. Determine which niche to focus on

There are worm farming niches you can take into consideration if you want to enter the industry. Here are some things to think about:

- Vermicomposting
- Worm casting
- Production of Worm Tea
- Worm Recycling of Green Waste

The level of industry competition

Although it has been a continuous endeavor, worm farming has been present since the early 1970s. Even if there is an enormous advantage to be had in other areas of agriculture, particularly fishing and crop cultivation, the demand for worms and their by-products is growing. However, the presence of well-known brands and their tested quality over time have made the worm farming sector quite competitive. However, it does not follow that you cannot enter the market as a new start-up with the appropriate marketing plan and branding.

4. Recognize your market's primary competitors

Starting and sticking with the worm farming industry has resulted in the success of numerous brands over time. Here are some of the more well-known brands:

- The Worm Farm
- Silver Bait L.L.C.
- Hungry Bin
- Uncle Jim's Worm Farm
- Wisconsin Redworms
- International Vermicrobe
- Clark County Crawlers

Economic Analysis

If bait worms are the product, you must be careful to make sure the worms get bigger and can be sold. This means that input costs may be higher. Vermicomposting is frequently easier, and the ingredients should be cheap or free.

Profitability is determined by the price paid for the final good after deducting inputs, labor, and capital expenditures. Every business environment will have a different set of these. You must conduct research and create a business plan as a result. Business plans can be very helpful in acquiring start-up capital and will go a long way toward guaranteeing a profitable business in the future.

5. Decide whether to start from scratch or buy a franchise

At most, you can start this business from scratch in your backyard, kitchen, or even basement. You can begin without no start-up cost using supplies and food from your own home. It might not be the best idea to start a huge business by purchasing a franchise when you can start small and learn as you go.

Purchasing farming equipment from an established worm farm that has a plan to purchase your products once they are produced is another start-up option you can take advantage of. The drawback to this strategy is how expensive the equipment is. It still comes down to building your farm from scratch rather than spending a lot of money when you first start out when you can still do so with little to no cost.

6. Be aware of any potential risks and difficulties you may encounter

Starting a worm farm can be difficult for many people since controlling the squirming critters while guarding against sickness is difficult. Many people believe they can't manage worms because they're nasty and strange, but despite how repulsive they appear, they're a simple way to improve your financial situation.

Another issue that many worm farmers deal with is the market for their products; they are unsure of the buyers or whom to sell their worms to when they are ready to be sold. Approaching purchasers and reaching an arrangement with them before maturity can resolve all of this.

7. Select the most appropriate legal entity (LLC, C Corp, S Corp)

You have the choice of selecting a general partnership, a limited partnership, an LLC, a "C" corporation, or an "S" corporation as the legal entity for your worm farm. It's important to say that each type of legal structure for a business has its own pros and cons, so you should think carefully about your options before deciding what type of legal structure to use for your worm farm.

It is recommended that you launch your farm business as a sole proprietorship or partnership and expand from there. You can make money at this level without being very ambitious or having a sizable amount of worms.

8. Choose a catchy company name

If you're considering starting a worm farming business, it's crucial to choose a name for the company that will immediately capture people's attention. Once they interact with you, this will pique their curiosity and cause them to patronize you. Several of the names you could use include;

- Vermi-Cycle

- Worm Warfare
- ReWorm Cycle
- Worm Paradise
- The Worm Hole
- WormInn
- Worm Zone
- BioWorm
- Zone of the Wormers
- Worm Up

9. Consult an agent to determine the best insurance options for you

Although there isn't a specialized insurance policy for worm farms and agricultural insurance isn't very prevalent, you can still insure other components of your worm farm, especially if you run it on a large scale. Below are some of the insurance coverage you can consider;

- Public and employer's liability Insurance
- Waste Liability Insurance
- Plant and machinery Insurance
- Combined Liability Insurance

- Employees Health Insurance
- Material Damage Insurance
- Workers Compensation
- Legal Expenses Insurance

10. Use trademarks, copyrights, and patents to safeguard your intellectual property

You should think about applying for intellectual property protection while beginning a worm farm. In addition to safeguarding your company's name and other papers, filing for intellectual property protection includes protecting your ideas and, of course, the logo you use.

You must start the procedure by submitting an application to the USPTO if you wish to register your trademark in the United States and file for intellectual property protection. As required by the USPTO, the ultimate approval of your trademark is subject to legal scrutiny.

11. Acquire the necessary professional certification

It's possible that you wouldn't typically need a professional certification to start a worm farm.

However, you might wish to do an agricultural course if you want to work at a very professional level.

12. Obtain the legal documents required to run your company

Worm farming is one of the unregulated sectors in the United States of America and, of course, around the world. However, you are expected to comply with the legal document requirements outlined in your country's constitution as a company and business enterprise. It's crucial to keep in mind that you only require these documents if your business is medium-scale or large-scale. Below are a few of the fundamental legal documents you must have in place before starting a worm farm.

- Certificate of Incorporation
- Business Plan
- Insurance Policy
- Non – disclosure Agreement
- Business License
- Memorandum of Understanding (MoU)
- Employment Agreement (offer letters)
- Company Bye-laws

- Operating Agreement
- Apostille
- Operating Agreement for LLCs

13. Write a business plan

One of the initial actions that you are expected to take when preparing to float a worm farm is to consult industry professionals to assist you in creating a strong and practical business strategy. The fact is that you would need a solid business strategy in place in order to successfully run a worm farm.

A workable business plan will help you avoid the trial-and-error method of running a firm, which is the blueprint required to manage a business successfully. You'll be able to run your company purposefully and possibly precisely; you'll know what to do, when, and how to deal with growth, problems, and expansion.

So, if you want to start a worm farm, you should make a thorough business plan that can pass a reality check. Your plan should include facts, figures, and other metrics that are specific to the country where you want to start your business.

The purpose of a business plan is to provide detailed instructions on how to operate your company successfully from the beginning, not just to have a business document in place. The management and growth plans for your worm farm should be covered in your business strategy. When placing numbers on revenue and earnings, etc., the general idea is to try as much as you can to be realistic and never to over-project. Writing a business plan is actually safer when underestimating, as you won't be as disappointed when you face reality.

At least two purposes are served by your business plan and budget projections:

- To determine your capacity to create a successful business.
- Trying to persuade a bank or other potential lender that you are sincere and have done your research. To create and run a successful firm, it is essential to keep thorough financial records and devote time to their analysis. If you go into every situation assuming there are ways to be more efficient or spend less, you will avoid unpleasant financial surprises.

14. Create a thorough cost analysis

The cost of starting a worm farm is determined by your start-up strategy; you may wish to purchase a franchise, an already-built tiny worm farm, or build one from scratch using readily available supplies. In any case, the following expenses are anticipated when taking on a floating worm form:

- The entire incorporation cost in the United States of America is $700.
- The estimated cost of insurance, licenses, and permits is $2,000.00.
- Launching an official website will cost you $700.
- Your first batch of worms will set you back $1,500.
- The cost of purchasing your farm equipment is $3,000.
- The additional expenditure ($2,500) includes business cards, signage, advertisements, and promotions.
- $1,000 for other miscellaneous expenses

A small-scale worm farm can be started for less than $5,000 based on the research-derived costs that are listed above. When viewed on a medium-sized screen, it can reach up to $12,000. You should aim to invest up to $25,000 on a broad basis.

15. Raise the required star-tup capital

A worm farmer needs a substantial financial investment to launch and maintain this business for a while before it starts to bring in enough money to support itself. In the United States of America, the majority of neighborhood banks might be open to financing this kind of enterprise.

There are still a number of additional options to raise money to support a worm farming business, such as;

- Individual savings
- Obtaining funds from family and friends
- Borrowing money from a wealthy angel investor
- Finding someone to partner with

16. Select an appropriate location for your business

Worms are seasonal species. Keep your worms cool, protected, and out of the sun during the summer. Move them to a sunny location over the winter to ensure productivity when there is a drop in temperature. Place it as close to your kitchen where possible to make it simple to maintain, and add scraps.

Choose an appropriate location for your worms, either inside or outside that can provide a lot of shade to combat the scorching summers and warmth to combat the exceptionally cold winters. Worms prefer temperatures between 55 and 75 degrees Fahrenheit. Your worms may also be impacted by local variations in soil salinity, temperature, and barometric pressure. A raised floor is preferable to ensure proper drainage, stop worms from escaping, and lower the chance of vermin attacks.

When deciding where to site your worm farm, keep the following in mind:

- Waste food accessibility
- Desired market reachability
- Presence of competitors in the geographic location
- The governing health legislation over the location
- The location's climatic and weather conditions

17. Hire people to meet your technical and human resource requirements

The amount of labor you will require for your business will depend on the size of your worm farm. If you are starting off at a lower level of output, then managing your farm will only require your direct input. However, as the worm farm grows over time, you will need to hire more staff.

When employing staff, keep in mind that you are managing a farm, such as a poultry or pig farm. Therefore, look for individuals who are enthusiastic and passionate about agriculture and animal husbandry.

18. Create a marketing strategy plan filled with ideas

Marketing would likely present the biggest barrier, as it does for most newly established businesses. The ideal scenario would be to attempt to convince nearby plant nurseries to sell your units or products, either directly or on consignment, and then pair this strategy with adverts on community notice boards, on the internet, or in the local publications.

Many vermiculture business owners have a stall at neighborhood fairs or flea markets where they demonstrate the process using an active worm kit, collect orders for worm farms and worms, and sell the vermicompost and worm tea to passersby. There is no doubt that starting will never be simple. But be patient; develop a memorable brand for yourself, offer a decent product at a reasonable price, and as people start to hear about you, things will undoubtedly get simpler.

The methods described below are examples you can use to promote your products and services.

- You can talk about your business in reputable business magazines, on TV and radio, on talk shows, and in interactive sessions about sustainable products.
- Add your company to local directories or the yellow pages (both online and offline).
- Utilize the internet to promote your company.
- Join your state's local chambers of commerce and agriculture.
- Promptness in contract bids for your product's supply

- Create various product packages for various client segments to accommodate their varying spending limits.
- Visit agricultural expositions and fairs.

19. Establish a reasonable price for your products and services

There isn't much to say here other than to make sure you get the opinion of other worm growers and worm suppliers when trying to set your price. This is necessary to enable you to determine a reasonable price for your products and services from their own pricing strategy.

20. Create steadfast competitive strategies that will help you succeed

You must be ready to proudly display your stock to customers. In order to count, examine, and observe your worms without dropping anything on the countertop, most businesses provide a convenient little container called a worm checker that you may empty your worms into. The quality of your worm stock must excite you to have it shared with your clients.

Without a doubt, this will be the best promotion you can do for your company. And the majority of worm growers are not prepared to implement this. By doing this, you will defeat your rivals. Any product will always be sold if people believe in it.

21. Come up with some ideas for keeping clients and customers happy

One of the simplest methods to boost customer retention and potentially attract new consumers in business, regardless of the industry you choose to set up shop in, is to always satisfy your customers. If your customers are happy with the services you provide, they are unlikely to look for another service provider or product.

According to statistics, a reduction in quality is one of the main causes for customers to seek out an alternative service or product. Customer service issues are another factor. You won't have trouble keeping your devoted customers if you can keep raising the caliber of your product and service offerings and the way you deliver customer service.

You can easily offer new items and prices to them, congratulate them on their birthdays and other anniversaries, monitor their progress, send mass text messages and tailored e-mails to them, and most importantly, you can quickly get their feedback and complaints.

22. Create a corporate identity and brand awareness strategies

Working with your advisors to develop publicity and advertising plans that will assist you in reaching the core of your target market is very crucial. First and foremost, you must make sure that your brand is well known and disseminated, which is why you must plan to collaborate with various social classes. All of the PR agencies in the sector should create all of your promotional materials and jingles.

The following are ways to strengthen your brand:

- Send introductory letters about your company along with your brochure to all the corporate organizations, local authorities, educational institutions, healthcare facilities, lodging establishments, and government offices in your target location.

- Promote your business in reputable magazines and websites.
- Add your company to local directories or the yellow pages (both online and offline).
- On satellite TV and radio stations, advertise your company.
- Social networking platforms and your official websites are effective ways to reach customers and market your business online.
- Make sure all of your employees wear branded clothing during work hours and brand all of your company's official vans, trucks, cars, and other vehicles.

23. Build a distributive / suppliers' network

With each batch of production, you are now producing high-quality products, and you will never be without suppliers or distributors. They will undoubtedly grin as they head to the bank, making them willing to set up camp with you rather than someone else.

A consistent supply of resources is also required for a business that relies heavily on raw materials that end users use. Regular product purchases from the supplier will ensure their steady supply to you.

Conclusion

Excessive waste materials have tremendous harmful effects on the earth and soil. More than one-third of our food ends up wasted in the trash. The idea of converting this waste into productive use and saving the earth should be a motivation for starting a worm farm. This also has the potential of generating a good amount of money when done right.

Worm farming is a major organic waste management process. It is almost free to start, could be done anywhere, especially at home, requires a very low cost of maintenance, and yet produces high-demand products. Through worm farming, you can create a high-value product out of waste, grow healthier and more beautiful plants, create highly sought-after fertilizer and save on gardening and fishing materials.

From the foregoing, worm farming is a very productive and lucrative business to venture into. Just anyone can go into worm farming. You can buy a franchise, but it is more advisable to start from scratch and learn more as you grow. As with every new business, there may be difficulties setting up your own worm farm, but virtually everything you need to get the ball rolling has been discussed in the pages of this book. So, ensure to digest all the information shared, then proceed to start your worm farm business right away!

I wish you all the best!